JOHN FLAMSTEED

GREAT ASTRONOMERS

ROBERT STAWELL BALL

GIFT *Certificate*

TO:

FROM:

DATE:

JOHN FLAMSTEED

GREAT ASTRONOMERS

BY

ROBERT STAWELL BALL

D.Sc, LL.D., F.R.S.

(July 1, 1840 – November 25, 1913)

LOWNDEAN PROFESSOR OF ASTRONOMY AND GEOMETRY
IN THE UNIVERSITY OF CAMBRIDGE

DIAMOND™
BOOKS

www.diamondbooks.ca

TORONTO, CANADA – 2017

DIAMOND BOOKS - CANADA

DIAMOND BOOKS

Toronto, ON, CANADA
http://www.diamondbooks.ca

BIBLIOGRAPHIC INFORMATION

'JOHN FLAMSTEED'
Is an extract or a chapter from the book
"GREAT ASTRONOMERS" by Robert Stawell Ball.

FIRST PUBLISHED IN: 1896.

PUBLISHED IN CANADA

Published in Canada by DIAMOND BOOKS - CANADA, an imprint of DIAMOND PUBLISHERS - http://www.diamondpublishers.com

FIRST EDITION: MARCH, 2017.

PAPERBACK EDITION : ISBN: 978-1-988357-50-8

PRINTED IN CANADA

PREFACE
TO THE ORIGINAL EDITION

It has been my object in these pages to present the life of each astronomer in such detail as to enable the reader to realise in some degree the man's character and surroundings; and I have endeavoured to indicate as clearly as circumstances would permit the main features of the discoveries by which he has become known.

There are many types of astronomers—from the stargazer who merely watches the heavens, to the abstract mathematician who merely works at his desk; it has, consequently, been necessary in the case of some lives to adopt a very different treatment from that which seemed suitable for others.

ROBERT STAWELL BALL.

The Observatory, Cambridge.
October, 1895.

NOTE: The full-length of 'Preface' can be found in its original edition of the book "GREAT ASTRONOMERS".

INTRODUCTION

Of all the natural sciences there is not one which offers such sublime objects to the attention of the inquirer as does the science of astronomy. From the earliest ages the study of the stars has exercised the same fascination as it possesses at the present day. Among the most primitive peoples, the movements of the sun, the moon, and the stars commanded attention from their supposed influence on human affairs. The practical utilities of astronomy were also obvious in primeval times. Maxims of extreme antiquity show how the avocations of the husbandman are to be guided by the movements of the heavenly bodies. The positions of the stars indicated the time to plough, and the time to sow. To the mariner who was seeking a way across the trackless ocean, the heavenly bodies .offered the only reliable marks by which his path could be guided. There was, accordingly, a stimulus both from intellectual curiosity and from practical necessity to follow the movements of the stars. Thus began a search for the causes of the ever-varying phenomena which the heavens display.

- ROBERT STAWELL BALL

Many of the earliest discoveries are indeed prehistoric. The great diurnal movement of the heavens, and the annual revolution of the sun, seem to have been known in times far more ancient than those to which any human monuments can be referred. The acuteness of the early observers enabled them to single out the more important of the wanderers which we now call planets. They saw that the star-like objects, Jupiter, Saturn, and Mars, with the more conspicuous Venus, constituted a class of bodies wholly distinct from the fixed stars among which their movements lay, and to which they bear such a superficial resemblance. But the penetration of the early astronomers went even further, for they recognized that Mercury also belongs to the same group, though this particular object is seen so rarely. It would seem that eclipses and other phenomena were observed at Babylon from a very remote period, while the most ancient records of celestial observations that we possess are to be found in the Chinese annals. The study of astronomy, in the sense in which we understand the word, may be said to have commenced under the reign of the Ptolemies at Alexandria. The most famous name in the science of this period is that of Hipparchus, who lived and worked at Rhodes about the year 160 B.C. It was his splendid investigations that first wrought the observed facts into a coherent branch of knowledge. He recognized the primary obligation which lies on the student of the heavens to compile as complete an inventory as possible of the objects which are there to be found. Hipparchus accordingly commenced by undertaking, on a small scale, a task exactly similar to that on which modern astronomers, with all available appliances of meridian circles, and photographic telescopes, are constantly engaged at the present day. He compiled a catalogue of the principal fixed stars, which is of special value to astronomers, as being the earliest work of its kind which has been handed down. He also studied the movements of the sun and the moon, and framed theories to account for the incessant changes which he saw in progress. He found a much more difficult problem in his attempt to interpret satisfactorily the complicated movements of the planets. With the view of constructing a theory which should give some coherent account of the subject, he made many observations of the places of these wandering stars. How great were the advances which Hipparchus accomplished may be appreciated if we reflect that, as a preliminary task to his more purely astronomical labours, he had to invent that branch of mathematical science by which alone the problems he proposed could be solved. It was for this purpose that he devised the indispensable method of calculation which we now know so well as trigonometry. Without the aid rendered by this beautiful art it would have been impossible for any n ally important advance in astronomical calculation to have been effected.

But the discovery which shows, beyond all others, that Hipparchus possessed one of the master-minds of all time was the detection of that remarkable celestial movement known as the precession of the equinoxes. The inquiry which conducted to this discovery involved a most profound investigation, especially when it is remembered that in the days of Hipparchus the means of observation of the heavenly bodies were only of the rudest description, and the available observations of earlier dates were extremely scanty. We can but look with astonishment on the genius of the man who, in spite of such difficulties, was able to detect such a phenomenon as the precession, and to exhibit its actual magnitude. I shall endeavour to explain the nature of this singular celestial movement, for it may be said to offer the first instance in the history of science in which we find that combination of accurate observation with skilful interpretation, of which, in the subsequent development of astronomy, we have so many splendid examples.

The word equinox implies the condition that the night is equal to the day. To a resident on the equator the night is no doubt equal to the day at all times in the year, but to one who lives on any other part of the earth, in either hemisphere, the night and the day are not generally equal. There is, however, one occasion in spring, and another in autumn, on which the day and the night are each twelve hours at all places on the earth. When the night and day are equal in spring, the point which the sun occupies on. the heavens is termed the vernal equinox. There is similarly another point in which the sun is situated at the time of the autumnal equinox. In any investigation of the celestial movements the positions of these two equinoxes on the heavens are of primary importance, and Hipparchus, with the instinct of genius, perceived their significance, and commenced to study them. It will be understood that we can always define the position of a point on the sky with reference to the surrounding stars. No doubt we do not see the stars near the sun when the sun is shining, but they are there nevertheless. The ingenuity of Hipparchus enabled him to determine the positions of each of the two equinoxes relatively to the stars which lie in its immediate vicinity. After examination of the celestial places of these points at different periods, he was led to the conclusion that each equinox was moving relatively to the stars, though that movement was so slow that twenty five thousand years would necessarily elapse before a complete circuit of the heavens was accomplished. Hipparchus traced out this phenomenon, and established it on an impregnable basis, so that all astronomers have ever since recognised the precession of the equinoxes) as one of the fundamental facts of astronomy. Not until nearly two thousand years after Hipparchus had made this splendid discovery was the explanation of its cause given by Newton.

From the days of Hipparchus down to the present hour the science of astronomy has steadily grown. One great observer after another has appeared from time to time, to reveal some new phenomenon with regard to the celestial bodies or their movements, while from time to time one commanding intellect after another has arisen to explain the true import of the facts of observations. The history of astronomy thus becomes inseparable from the history of the great men to whose labours its development is due. In the ensuing chapters we have endeavoured to sketch the lives and the work of the great philosophers, by whose labours the science of astronomy has been created.

We shall commence with Ptolemy, who, after the foundations of the science had been laid by Hipparchus, gave to astronomy the form in which it was taught throughout the Middle Ages. We shall next see the mighty revolution in our conceptions of the universe which are associated with the name of Copernicus. We then pass to those periods illumined by the genius of Galileo and Newton, and afterwards we shall trace the careers of other more recent discoverers, by whose industry and genius the boundaries of human knowledge have been so greatly extended. Our history will be brought down late enough to include some of the illustrious astronomers who laboured in the generation which has just passed away.

- ROBERT STAWELL BALL

TABLE OF CONTENTS

CHAPTERS

1. **Chapter # 7***
John Flamsteed 1

2. A Brief Biographical Sketch 27
- *Complimentary Material*

ABOUT THE AUTHOR 45

**An original Chapter from the book: "GREAT ASTRONOMERS", by Robert Stawell Ball.*

- ROBERT STAWELL BALL

CHAPTER 7

GREAT ASTRONOMERS

John Flamsteed

(August 19, 1646 – December 31, 1719)

- ROBERT STAWELL BALL

John Flamsteed

(August 19, 1646 – December 31, 1719)

CHAPTER 7

GREAT ASTRONOMERS: JOHN FLAMSTEED

AMONG the manuscripts preserved at Greenwich Observatory are certain documents in which Flamsteed gives an account of his own life. We may commence our sketch by quoting the following passage from this autobiography:

"To keep myself from idleness, and to recreate myself, I have intended here to give some account of my life, in my youth, before the actions thereof, and the providences of God therein, be too far passed out of my memory; and to observe the accidents of all my years, and inclinations of my mind, that whosoever may light upon these papers may see I was not so wholly taken up, either with my father's business or my mathematics, but that I both admitted and found time for other as weighty considerations."

The chief interest which attaches to the name of Flamsteed arises from the fact that he was the first of the illustrious series of Astronomers Royal who have presided over Greenwich Observatory. In that capacity Flamsteed was able to render material assistance to Newton by providing him with the observations which his lunar theory required.

John Flamsteed was born at Denby, in Derbyshire, on the 19th of August, 1646. His mother died when he was three years old, and the second wife, whom his father took three years later, only lived until Flamsteed was eight, there being also two younger sisters. In his boyhood the future astronomer tells us that he was very fond of those romances which affect boy's imagination, but as he writes,

"At twelve years of age I left all the wild ones and betook myself to read the better sort of them, which, though they were not probable, yet carried no seeming impossibility in the picturing."

By the time Flamsteed was fifteen years old he had embarked in still more serious work, for he had read Plutarch's "Lives," Tacitus' "Roman History," and many other books of a similar description. In 1661 he became ill with some serious rheumatic affection, which obliged him to be withdrawn from school. It was then for the first time that he received the rudiments of a scientific education. He had, however, attained his sixteenth year before he made any progress in arithmetic. He tells us how his father taught him "the doctrine of fractions," and "the golden rule of three"—lessons which he seemed to have learned easily and quickly. One of the books which he read at this time directed his attention to astronomical instruments, and he was thus led to construct for himself a quadrant, by which he could take some simple astronomical observations. He further calculated a table to give the sun's altitudes at different hours, and thus displayed those tastes for practical astronomy which he lived to develop so greatly. It appears that these scientific studies were discountenanced by his father, who designed that his son should follow a business career.

Flamsteed's natural inclination, however, forced him to prosecute astronomical work, notwithstanding the impediments that lay in his path. Unfortunately, his constitutional delicacy seems to have increased, and he had just completed his eighteenth year, "when," to use his own words, "the winter came on and thrust me again into the chimney, whence the heat and the dryness of the preceding summer had happily once before withdrawn me. But, it not being a fit season for physic, it was thought fit to let me alone this winter, and try the skill of another physician on me in the spring."

It appears that at this time a quack, named Valentine Greatrackes, was reputed to have effected most astonishing cures in Ireland merely by the stroke of his hands, without the application of any medicine whatever. Flamsteed's father, despairing of any remedy for his son from the legitimate branch of the profession, despatched him to Ireland on August 26th, 1665, he being then, as recorded with astronomical accuracy, "nineteen years, six days, and eleven hours old." The young astronomer, accompanied by a friend, arrived on a Tuesday at Liverpool, but the wind not being favourable, they remained there till the following Friday, when a shift of the wind to the east took place. They embarked accordingly on a vessel called the *Supply* at noon, and on Saturday night came in sight of Dublin. Ere they could land, however, they were nearly being wrecked on Lambay Island. This peril safely passed, there was a long delay for quarantine before they were at last allowed on shore. On Thursday, September 6th, they set out from Dublin, where they had been sojourning at the "Ship" Hotel, in Dame Street, towards Assaune, where Greatrackes received his patients.

Flamsteed gives an interesting account of his travels in Ireland. They dined at Naas on the first day, and on September 8th they reached Carlow, a town which is described as one of the fairest they saw on their journey. By Sunday morning, September 10th, having lost their way several times, they reached Castleton, called commonly Four Mile Waters. Flamsteed inquired of the host in the inn where they might find a church, but was told that the minister lived twelve miles away, and that they had no sermon except when he came to receive his tithes once a year, and a woman added that, "they had plenty enough of everything necessary except the word of God." The travellers accordingly went on to Cappoquin, which lies up the river Blackwater, on the road to Lismore, eight miles from Youghal. Thence they immediately started on foot to Assaune, about a mile from Cappoquin, and entering into the house of Mr. Greatrackes, they saw him touch several patients, "whereof some were nearly cured, others were on the mending hand, and some on whom his strokes had no effect." Flamsteed was touched by the famous quack on the afternoon of September 11th, but we are hardly surprised to hear his remark that "he found not his disease to stir." Next morning the astronomer came again to see Mr. Greatrackes, who had "a kind of majestical yet affable presence, and a composed carriage." Even after the third touching had been submitted to, no benefit seems to have been derived. We must, however, record, to the credit of Mr. Greatrackes, that he refused to accept any payment from Flamsteed, because he was a stranger.

- ROBERT STAWELL BALL

Finding it useless to protract his stay any longer, Flamsteed and his friend set out on their return to Dublin. In the course of his journey he seems to have been much impressed with Clonmel, which he describes as an "exceedingly pleasantly seated town."

Flamsteed's House.

But in those days a journey to Ireland was so serious an enterprise that when Flamsteed did arrive safely back at Derby after an absence of a month, he adds, "For God's providence in this journey, His name be praised, Amen."

As to the expected benefits to his health from the expedition we may quote his own words:

"In the winter following I was indifferent hearty, and my disease was not so violent as it used to be at that time formerly. But whether through God's mercy I received this through Mr. Greatrackes' touch, or my journey and vomiting at sea, I am uncertain; but, by some circumstances, I guess that I received a benefit from both."

It is evident that by this time Flamsteed's interest in all astronomical matters had greatly increased. He studied the construction of sun-dials, he formed a catalogue of seventy of the fixed stars, with their places on the heavens, and he computed the circumstances of the solar eclipse which was to happen on June 22nd, 1666. It is interesting to note that even in those days the doctrines of the astrologers still found a considerable degree of credence, and Flamsteed spent a good deal of his time in astrological studies and computations. He investigated the methods of casting a nativity, but a suspicion, or, indeed, rather more than a suspicion, seems to have crossed his mind as to the value of these astrological predictions, for he says in time, “I found astrology to give generally strong conjectural hints, not perfect declarations.”

All this time, however, the future Astronomer Royal was steadily advancing in astronomical inquiries of a recondite nature. He had investigated the obliquity of the ecliptic with extreme care, so far as the circumstances of astronomical observation would at that time permit. He had also sought to discover the sun's distance from the earth in so far as it could be obtained by determining when the moon was exactly half illuminated, and he had measured, with much accuracy, the length of the tropical year. It will thus be seen that, even at the age of twenty, Flamsteed had made marked progress, considering how much his time had been interfered with by ill-health.

Other branches of astronomy began also to claim his attention. We learn that in 1669 and 1670 he compared the planets Jupiter and Mars with certain fixed stars near which they passed. His instrumental means, though very imperfect, were still sufficient to enable him to measure the intervals on the celestial sphere between the planets and the stars. As the places of the stars were known, Flamsteed was thus able to obtain the places of the planets. This is substantially the way in which astronomers of the present day still proceed when they desire to determine the places of the planets, inasmuch as, directly or indirectly, those places are always obtained relatively to the fixed stars. By his observations at this early period, Flamsteed was, it is true, not able to obtain any very great degree of accuracy; he succeeded, however, in proving that the tables by which the places of the planets were ordinarily given were not to be relied upon.

Flamsteed's labours in astronomy and in the allied branches of science were now becoming generally known, and he gradually came to correspond with many distinguished men of learning. One of the first occasions which brought the talents of the young astronomer into fame was the publication of some calculations concerning certain astronomical phenomena which were to happen in the year 1670. In the monthly revolution of the moon its disc passes over those stars which lie along its track. The disappearance of a star by the interposition of the moon is called an "occultation." Owing to the fact that our satellite is comparatively near us, the position which the moon appears to occupy on the heavens varies from different parts of the earth. It consequently happens that a star which would be occulted to an observer in one locality, would often not be occulted to an observer who was situated elsewhere. Even when an occultation is visible from both places, the times at which the star disappears from view will, generally speaking, be different. Much calculation is therefore necessary to decide the circumstances under which the occultations of stars may be visible from any particular station.

Having a taste for such computations, Flamsteed calculated the occultations which were to happen in the year 1670, it being the case that several remarkable stars would be passed over by the moon during this year. Of course at the present time we find such information duly set forth in the *Nautical Almanac*, but a couple of centuries ago there was no such source of astronomical knowledge as is now to be found in that invaluable publication, which astronomers and navigators know so well. Flamsteed accordingly sent the results of his work to the President of the Royal Society. The paper which contained them was received very favourably, and at once brought Flamsteed into notice among the most eminent members of that illustrious body, one of whom, Mr. Collins, became through life his faithful friend and constant correspondent. Flamsteed's father was naturally gratified with the remarkable notice which his son was receiving from the great and learned; accordingly he desired him to go to London, that he might make the personal acquaintance of those scientific friends whom he had only known by correspondence previously. Flamsteed was indeed glad to avail himself of this opportunity. Thus he became acquainted with Dr. Barrow, and especially with Newton, who was then Lucasian Professor of Mathematics at Cambridge. It seems to have been in consequence of this visit to London that Fiamsteed entered himself as a member of Jesus College, Cambridge. We have but little information as to his University career, but at all events he took his degree of M.A. on June 5th, 1674.

Up to this time it would seem that Fiamsteed had been engaged, to a certain extent, in the business carried on by his father. It is true that he does not give any explicit details, yet there are frequent references to journeys which he had to take on business matters. But the time now approached when Flamsteed was to start on an independent career, and it appears that he took his degree in Cambridge with the object of entering into holy orders, so that he might settle in a small living near Derby, which was in the gift of a friend of his father, and would be at the disposal of the young astronomer. This scheme was, however, not carried out, but Flamsteed does not tell us why it failed, his only remark being, that "the good providence of God that had designed me for another station ordered it otherwise."

Sir Jonas Moore, one of the influential friends whom Flamsteed's talents had attracted, seems to have procured for him the position of king's astronomer, with a salary of £100 per annum. A larger salary appears to have been designed at first for this office, which was now being newly created, but as Flamsteed was resolved on taking holy orders, a lesser salary was in his case deemed sufficient. The building of the observatory, in which the first Astronomer Royal was to be installed, seems to have been brought about, or, at all events, its progress was accelerated, in a somewhat curious manner.

A Frenchman, named Le Sieur de S. Pierre, came over to London to promulgate a scheme for discovering longitudes, then a question of much importance. He brought with him introductions to distinguished people, and his mission attracted a great deal of attention. The proposals which he made came under Flamsteed's notice, who pointed out that the Frenchman's projects were quite inapplicable in the present state of astronomical science, inasmuch as the places of the stars were not known with the degree of accuracy which would be necessary if such methods were to be rendered available. Flamsteed then goes on to say:— "I heard no more of the Frenchman after this; but was told that my letters had been shown King Charles. He was startled at the assertion of the fixed stars' places being false in the catalogue, and said, with some vehemence, he must have them anew observed, examined, and corrected, for the use of his seamen."

The first question to be settled was the site for the new observatory. Hyde Park and Chelsea College were both mentioned as suitable localities, but, at Sir Christopher Wren's suggestion, Greenwich Hill was finally resolved upon. The king made a grant of five hundred pounds of money. He gave bricks from Tilbury Fort, while materials, in the shape of wood, iron, and lead, were available from a gatehouse demolished in the Tower. The king also promised whatever further material aid might be shown to be necessary. The first stone of the Royal Observatory was laid on August 10th, 1675, and within a few years a building was erected in which the art of modern practical astronomy was to be created. Flamsteed strove with extraordinary diligence, and in spite of many difficulties, to obtain a due provision of astronomical instruments, and to arrange for the carrying on of his observations. Notwithstanding the king's promises, the astronomer was, however, but scantily provided with means, and he had no assistants to help him in his work. It follows that all the observations, as well as the reductions, and, indeed, all the incidental work of the observatory, had to be carried on by himself alone. Flamsteed, as we have seen, had, however, many staunch friends. Sir Jonas Moore in particular at all times rendered him most valuable assistance, and encouraged him by the warm sympathy and keen interest which he showed in astronomy. The work of the first Astronomer Royal was frequently interrupted by recurrent attacks of the complaints to which we have already referred. He says himself that “his distempers stick so close that he cannot remove them, “and he lost much time by prostration from headaches, as well as from more serious affections.

The year 1678 found him in the full tide of work in his observatory. He was specially engaged on the problem of the earth's motion, which he sought to derive from observations of the sun and of Venus. But this, as well as many other astronomical researches which he undertook, were only subsidiary to that which he made the main task of his life, namely, the formation of a catalogue of fixed stars. At the time when Flamsteed commenced his career, the only available catalogue of fixed stars was that of Tycho Brahe. This work had been published at the commencement of the seventeenth century, and it contained about a thousand stars. The positions assigned to these stars, though obtained with wonderful skill, considering the many difficulties under which Tycho laboured, were quite inaccurate when judged by our modern standards. Tycho's instruments were necessarily most rudely divided, and he had, of course, no telescopes to aid him. Consequently it was merely by a process of sighting that he could obtain the places of the stars. It must further be remembered that Tycho had no clocks, and no micrometers. He had, indeed, but little correct knowledge of the motions of the heavenly bodies to guide him. To determine the longitudes of a few principal stars he conceived the ingenious idea of measuring by day the position of Venus with respect to the sun, an observation which the exceptional brightness of this planet rendered possible without telescopic aid, and then by night he observed the position of Venus with regard to the stars.

It has been well remarked by Mr. Baily, in his introduction to the "British Catalogue of Stars," that "Flamsteed's observations, by a fortunate combination of circumstances, commenced a new and a brilliant era. It happened that, at that period, the powerful mind of Newton was directed to this subject; a friendly intercourse then existed between these two distinguished characters; and thus the first observations that could lay any claim to accuracy were at once brought in aid of those deep researches in which our illustrious geometer was then engaged. The first edition of the 'Principia' bears testimony to the assistance afforded by Flamsteed to Newton in these inquiries; although the former considers that the acknowledgment is not so ample as it ought to have been."

Although Flamsteed's observations can hardly be said to possess the accuracy of those made in more recent times, when instruments so much superior to his have been available, yet they possess an interest of a special kind from their very antiquity. This circumstance renders them of particular importance to the astronomer, inasmuch as they are calculated to throw light on the proper motions of the stars. Flamsteed's work may, indeed, be regarded as the origin of all subsequent catalogues, and the nomenclature which he adopted, though in some respects it can hardly be said to be very defensible, is, nevertheless, that which has been adopted by all subsequent astronomers. There were also a great many errors, as might be expected in a work of such extent, composed almost entirely of numerical detail. Many of these errors have been corrected by

Baily himself, the assiduous editor of “Flamsteed's Life and Works,” for Flamsteed was so harassed from various causes in the latter part of his life, and was so subject to infirmities all through his career, that he was unable to revise his computations with the care that would have been necessary. Indeed, he observed many additional stars which he never included in the British Catalogue. It is, as Baily well remarks, "rather a matter of astonishment that he accomplished so much, considering his slender means, his weak frame, and the vexations which he constantly experienced."

Flamsteed had the misfortune, in the latter part of his life, to become estranged from his most eminent scientific contemporaries. He had supplied Newton with places of the moon, at the urgent solicitation of the author of the "Principia," in order that the lunar theory should be carefully compared with observation. But Flamsteed appears to have thought that in Newton's further request for similar information, he appeared to be demanding as a right that which Flamsteed considered he was only called upon to render as a favour. A considerable dispute grew out of this matter, and there are many letters and documents, bearing on the difficulties which subsequently arose, that are not, perhaps, very creditable to either party.

Notwithstanding his feeble constitution, Flamsteed lived to the age of seventy-three, his death occurring on the last day of the year 1719.

COMPLIMENTARY MATERIAL

A BRIEF BIOGRAPHICAL SKETCH

JOHN FLAMSTEED

- ROBERT STAWELL BALL

John Flamsteed
(August 19, 1646 – December 31, 1719)

COMPLIMENTARY MATERIAL

A BRIEF BIOGRAPHICAL SKETCH

JOHN FLAMSTEED

John Flamsteed FRS (August 19, 1646 - December 31, 1719) was an English astronomer and the first Astronomer Royal. He catalogued over 3000 stars.

Life

Flamsteed was born in Denby, Derbyshire, England, the only son of Stephen Flamsteed and his first wife, Mary Spadman. He was educated at the free school of Derby, and was educated at Derby School, in St Peter's Churchyard, Derby, near where his father carried on a malting business. At that time, most masters of the school were Puritans. Flamsteed had a solid knowledge of Latin, essential for reading the scientific literature of the day, and a love of history, leaving the school in May, 1662.[2]:3–4

His progress to Jesus College, Cambridge, recommended by the Master of Derby School, was delayed by some years of chronic ill health. During those years, Flamsteed gave his father some help in his business, and from his father learnt arithmetic and the use of fractions, developing a keen interest in mathematics and astronomy. In July 1662, he was fascinated by the thirteenth-century work of Johannes de Sacrobosco, De sphaera mundi, and on 12 September 1662 observed his first partial solar eclipse. Early in 1663, he read Thomas Fale's The Art of Dialling, which set off an interest in sundials. In the summer of 1663, he read Wingate's Canon, William Oughtred's Canon, and Thomas Stirrup's Art of Dialling. At about the same time, he acquired Thomas Street's Astronomia Carolina, or A New Theory of the Celestial Motions (Caroline Tables). He associated himself with local gentlemen interested in astronomy, including William Litchford, whose library included the work of the astrologer John Gadbury which included astronomical tables by Jeremiah Horrocks, who had died in 1641 at the age of twenty-two. Flamsteed was greatly impressed (as Isaac Newton had been) by the work of Horrocks.

In August 1665, at the age of nineteen and as a gift for his friend Litchford, Flamsteed wrote his first paper on astronomy, entitled Mathematical Essays, concerning the design, use and construction of an astronomer's quadrant, including tables for the latitude of Derby.

In September 1670, Flamsteed visited Cambridge and entered his name as an undergraduate at Jesus College. While it seems he never took up full residence, he was there for two months in 1674, and had the opportunity to hear Isaac Newton's Lucasian Lectures.

Ordained a deacon, he was preparing to take up a living in Derbyshire when he was invited to London by his patron Jonas Moore, Surveyor-General of the Ordnance. Moore had recently made an offer to the Royal Society to pay for the establishment of an observatory. These plans were, however, preempted when Charles II was persuaded by his mistress, Louise de Kérouaille, Duchess of Portsmouth, to hear about a proposal to find longitude by the position of the Moon from an individual known as Le Sieur de St Pierre. Charles appointed a Royal Commission to examine the proposal in December 1674, consisting of Lord Brouncker, Seth Ward, Samuel Moreland, Christopher Wren, Silius Titus,

John Pell and Robert Hooke.

Having arrived in London on 2 February 1675, and staying with Jonas Moore at the Tower of London, Flamsteed had the opportunity to be taken by Titus to meet the King. He was subsequently admitted as an official Assistant to the Royal Commission and supplied observations in order to test St Pierre's proposal and to offer his own comments. The Commission's conclusions were that, although St Pierre's proposal was not worth further consideration, the King should consider establishing an observatory and appointing an observer in order to better map the stars and the motions of the Moon in order to underpin the successful development of the lunar-distance method of finding longitude.

On 4 March 1675 Flamsteed was appointed by royal warrant "The King's Astronomical Observator" — the first English Astronomer Royal, with an allowance of £100 a year. The warrant stated his task as "rectifieing the Tables of the motions of the Heavens, and the places of the fixed stars, so as to find out the so much desired Longitude of places for Perfecteing the Art of Navigation".[5] In June 1675, another royal warrant provided for the founding of the Royal Greenwich Observatory, and Flamsteed laid the foundation stone on 10 August.

In February 1676, he was admitted a Fellow of the Royal Society, and in July, he moved into the Observatory where he lived until 1684, when he was finally appointed priest to the parish of Burstow, near Crawley in Surrey. He held that office, as well as that of Astronomer Royal, until his death. He is buried at Burstow, and the east window in the church was dedicated to him as a memorial.

The will of Flamsteed's widow, Margaret, left instructions for her own remains to be deposited "in the same Grave in which Mr John Flamsteed is buryed in the Chancell of Burstow Church." Intriguingly, she also left instructions, and twenty five pounds, for the executor of her will to place "in the aforesaid Chancell of Burstow… A Marble stone or Monument, with an inscription in

Latin, in memory of the late Reverend Mr. John Flamsteed." It seems no such monument was created, and almost 200 years later, a plaque was placed to mark his burial in the chancel.

Memorial marking grave of John Flamsteed and his wife in the Chancel of St Bartholomew, Burstow

After his death, his papers and scientific instruments were taken by his widow. The papers were returned many years later, but the instruments disappeared.

Scientific Work

Flamsteed accurately calculated the solar eclipses of 1666 and 1668. He was responsible for several of the earliest recorded sightings of the planet Uranus, which he mistook for a star and catalogued as '34 Tauri'. The first of these was in December 1690, which remains the earliest known sighting of Uranus by an astronomer.

On 16 August 1680 Flamsteed catalogued a star, 3 Cassiopeiae, that later astronomers were unable to corroborate. Three hundred years later, the American astronomical historian William Ashworth suggested that what Flamsteed may have seen was the most recent supernova in the galaxy's history, an event which would leave as its remnant the strongest radio source outside of the solar system, known in the third Cambridge (3C) catalogue as 3C 461 and commonly called Cassiopeia A by astronomers. Because the position of "3

Cassiopeiae" does not precisely match that of Cassiopeia A, and because the expansion wave associated with the explosion has been worked backward to the year 1667 and not 1680, some historians feel that all Flamsteed may have done was incorrectly note the position of a star already known.

In 1681 Flamsteed proposed that the two great comets observed in November and December 1680 were not separate bodies, but rather a single comet travelling first towards the Sun and then away from it. Although Isaac Newton first disagreed with Flamsteed, he later came to agree with him and theorized that comets, like planets, moved around the sun in large, closed elliptical orbits. An angry Flamsteed later learned that Newton had gained access to Flamsteed's observations and data, with the aid of Edmund Halley.

As Astronomer Royal, Flamsteed spent some forty years observing and making meticulous records for his star catalogue, which would eventually triple the number of entries in Tycho Brahe's sky atlas.

Unwilling to risk his reputation by releasing unverified data, he kept the incomplete records under seal at Greenwich. In 1712, Isaac Newton, then President of the Royal Society, and Edmund Halley again obtained Flamsteed's data and published a pirated star catalogue. Flamsteed managed to gather three hundred of the four hundred printings and burned them. "If Sir I.N. would be sensible of it, I have done both him and Dr. Halley a great kindness," he wrote to his assistant Abraham Sharp.

- ROBERT STAWELL BALL

Bust of John Flamsteed in the Museum
of the Royal Greenwich Observatory

In 1725 Flamsteed's own version of Historia Coelestis Britannica was published posthumously, edited by his wife, Margaret. This contained Flamsteed's observations, and included a catalogue of 2,935 stars to much greater accuracy than any prior work. It was considered the first significant contribution of the Greenwich Observatory, and the numerical Flamsteed designations for stars that were added subsequently to a French edition are still in use. In 1729 his wife published his Atlas Coelestis, assisted by Joseph Crosthwait and Abraham Sharp, who were responsible for the technical side.

Honours

- Fellow of the Royal Society (1677)
- The Flamsteed Astronomy Society is named in honour of the first Astronomer Royal and is based befittingly at The Royal Observatory Greenwich.
- The crater Flamsteed on the Moon is named after him.
- The asteroid (4987) Flamsteed is named in his honour.
- Numerous schools and colleges in Derbyshire have been named after him. The science block at John Port School is named Flamsteed in recognition of his work for science. John Flamsteed Community School in Denby carries his name.

- Flamsteed House at the Ecclesbourne School in Duffield is also named after him.

- In early 2013 Derby City Council and Derby Civic Society announced they would erect a Blue Plaque in his honour at the Queen Street former Clock Works in Derby.) (This has since been erected and also honours Joseph Wright of Derby who lived in the house formerly owned by Flamsteed.)

- ROBERT STAWELL BALL

ABOUT THE AUTHOR

ROBERT STAWELL BALL

(July 1, 1840 – November 25, 1913)

Sir Robert Stawell Ball was an Irish astronomer who founded the screw theory.

He was the son of naturalist Robert Ball and Amelia Gresley Hellicar. He was born in Dublin.

- ROBERT STAWELL BALL

Ball worked for Lord Rosse from 1865 to 1867. In 1867 he became Professor of Applied Mathematics at the Royal College of Science in Dublin. There he lectured on mechanics and published an elementary account of the science.

In 1874 Ball was appointed Royal Astronomer of Ireland and Andrews Professor of Astronomy in the University of Dublin at Dunsink Observatory.

Ball contributed to the science of kinematics by delineating the screw displacement:

When Ball and the screw theorists speak of screws they no longer mean actual cylindrical objects with helical threads cut into them but the possible motion of any body whatsoever, including that of the screw independently of the nut.

Ball's treatise *The Theory of Screws* (1876) is now in the public domain. His work on screw dynamics earned him in 1879 the

Cunningham Medal of the Royal Irish Academy.

In 1882 Popular Science Monthly carried his article "*A Glimpse through the Corridors of Time*". The following year it carried his two-part article on "*The Boundaries of Astronomy*".

Ball expounded the tides in *Time and Tide*: a Romance of the Moon In 1892 he was appointed Lowndean Professor of Astronomy and Geometry at Cambridge University at the same time becoming director of the Cambridge Observatory. He was a fellow of King's College, Cambridge.

In 1900 Cambridge University Press published *A Treatise on the Theory of Screws*. That year he also published *The Story of the Heavens*. Much in the limelight, he stood as President of the Quaternion Society. He was also President of the Mathematical Association in 1900.

In 1908 he published *A Treatise on Spherical Astronomy*, which is a textbook on astronomy starting from spherical trigonometry and the celestial sphere, considering atmospheric refraction and aberration of light, and introducing basic use of a generalised instrument.

His work *The Story of the Heavens* is mentioned in the "Ithaka" chapter of Ulysses. His lectures, articles and books (e.g. *Starland* and *The Story of the Heavens*) were mostly popular and simple in style.

He died in Cambridge and was buried at the Parish of the Ascension Burial Ground in Cambridge, with his wife Lady Francis Elizabeth Ball. Their children were: Frances Amelia, Robert Steele, William Valentine (later Sir), Mary Agnetta, Charles Rowan Hamilton, and Randall Gresley (later Colonel).

LECTURES

Ball became celebrated for his popular lectures on science. He gave an estimated 2500 lectures between 1875 and 1910 in towns and cities across Britain and Ireland. In 1892, 1898 and 1900 he was invited to deliver the Royal Institution Christmas Lecture. *Astronomy; Astronomy* and *Great Chapters from the Book of Nature.*

- ROBERT STAWELL BALL

NOTES

NOTES

NOTES

NOTES

NOTES

NOTES

Made in the USA
Monee, IL
07 November 2026

68170760R00048